Clay Anatomy

Exploring the Human Body with Artistic Minds

by Lucy Shi and Angela Liang

Clay Anatomy
Exploring the Human Body with Artistic Minds
by Lucy Shi and Angela Liang
1. SCI036000   SCIENCE / Life Sciences / Human Anatomy & Physiology
2. EDU057000   EDUCATION / Arts in Education
3. ART027000   ART / Study & Teaching

ISBN (paperback): 979-8-88636-057-8
ISBN (ebook): 979-8-88636-058-5

Library of Congress Control Number: 2025914297

Cover design by Lucy Shi, Angela Liang, and Lewis Agrell

Printed in the United States of America

Authority Publishing
13389 Folsom Blvd #300-256
Folsom, CA 95630
800-877-1097
www.AuthorityPublishing.com

Why you should read this book

Have you ever wondered what happens in your stomach when you eat food? Or what happens when you start bleeding because of a scraped knee? Many people aren't too fond of the idea of blood, but there are genuinely interesting and complex systems that explain how our bodies operate. This book uses a familiar material, clay, to explain how human organs operate, so people of all ages can learn about anatomy.

Clay Anatomy features:

> Life-sized models accompanied by scientific descriptions
> Fun cartoons that share advice for maintaining a healthy lifestyle
> Interesting facts about your body that you can share with friends and family
> Easy-to-understand visuals and text that is perfect for all ages

Clay Anatomy explains different organs' functions in a fun and engaging manner. Let's dive into this journey to learn about the basics of biology and anatomy together!

Foreword

What happens when curiosity meets creativity? Something extraordinary. In this delightful book, two young authors bring human biology to life by transforming clay into colorful, detailed models of the human body. With each page, they make anatomy fun, accessible, and unforgettable. This book is more than a guide to our amazing bodies—it's a celebration of curiosity and creativity. These young creators show us that science isn't just for textbooks; it's something we can explore with our hands, imaginations, and a sense of wonder.

Whether you're a budding scientist, an aspiring artist, or just curious, this book will leave you inspired to learn, create, and explore. Enjoy the journey through these two artists and scientists!

Yalda Afshar, MD, PhD
Maternal Fetal Medicine
UCLA Health

Anatomy atlases have been a staple of medical education for centuries, ever since the likes of Vesalius discovered the wonderful inner workings of the human body. From the Renaissance on, physicians and artists collaborated to make discoveries in anatomy, physiology, medicine, and surgery reproducible and shareable by all: they used their complementary skills and expertise to create intricate and precise works of art.

Printed illustrations of the skeletal system, the muscles, and the organs were sometimes supplemented with wax models – after all, most of us are visual learners, and the addition of tactile, didactic models that can be manipulated in space makes it all so much more vivid.

Thus, it is in a long-honored tradition that Lucy and Angela have created these wonderful two- and three-dimensional works of art to explain the various organ systems of our body. Mirroring the illustrator-physician dyad, our guides through this journey are the brain's yin and the heart's yang, the duality of art and science.

The illustrations and clay models are beautiful to look at and accessible to even the youngest of budding scientists – but don't let the simplicity of the images fool you: what the organs lack in microvasculature, neuronal networks, and intricate layers is amply compensated by a clear, accurate description of the principles of digestion, respiration, and other vital physiologic functions. It is a gorgeous stand-alone oeuvre – and it will stimulate the reader to seek out more knowledge.

Francois I. Luks, MD, PhD
Professor of Pediatric Surgery
Author and Medical Illustrator
Brown University

The human body is a reflection of both art and science. The design of our organs and the systems that work together to sustain life is a marvel. Lucy and her friend's ambitious attempt to explore the worlds of our bodies in artistic form is an appropriate reflection of the magic of life, and through their work, the unseen becomes visible and the mysterious becomes understood. Physically crafting anatomical features from clay offers a sense of their design, and this hands-on artistic approach invites readers to contemplate the makings of our bodies. The more we understand our own bodies, the better we can take care of them and appreciate the blessings of health.

Part artistic expression and part science manual, this book will engage audiences young and old, and illustrate the interdisciplinary connection between two subjects that are not often enough associated. As an educator and school leader, I revel in my students' abilities to make connections and transfer knowledge from one field to another. That is exactly what we see in this book. *Clay Anatomy* is a reflection of our students' curiosity, imagination, and vision. It showcases their passion and creativity in inspiring others to view science through an artistic lens. I am proud to witness my students grow as budding scientists and artists, and I have no doubt this book will spark wonder in many young minds.

Jon Wimbish
Head of Middle School
Harvard-Westlake School

Preface

It all started with a simple idea and a strong curiosity to learn more.

One summer day when we were eating ice cream, we came upon the topic of discussing what happens after swallowing each bite of food. While discussing the digestive system, we realized that most resources explaining how the body functions are filled with overly complex vocabulary and somewhat scary anatomical images. There aren't many books that break down the components of the human body in an easily digestible form for younger students. An idea sparked that maybe we could create anatomical models using clay to show everyone, even children, how amazing our bodies are. We had always loved playing with clay as children – shaping it into flowers, donuts, and anything else our imagination could dream of. What if we shaped and molded clay organs instead?

And so, the idea for *Clay Anatomy* was born.

With this in mind, we began to form clay stomachs, hearts, livers, and kidneys, and quickly discovered that learning about science could be as fun as play. We personally molded each organ, etched in small details, and painted them with vibrant colors. Each model contains our love and dedication toward biology — you can even see our fingerprints in the clay, connecting us to each of our readers.

This book is a result of a multitude of imaginative ideas, passion for art, and collaboration among science-lovers. Each page features a clay model paired with a simple explanation and helpful tips to help you stay healthy. We designed this with the mission to make learning about anatomy fun and accessible to everyone — whether you're a kid who loves to ask questions, a parent who loves to play along, or simply someone eager to learn.

We hope through *Clay Anatomy*, you'll discover the wonders of your own body in an intriguing and eager manner. We are thrilled to share this journey with you, and we're so grateful for the amazing support we've received from friends, family, and experts alike.

Thank you for joining us on this adventure. We can't wait for you to explore the world of *Clay Anatomy*!

Lucy and Angela

Acknowledgements

We had the pleasure of working with so many wonderful people on this adventure. We'd like to give a huge thank you to everyone who helped us and inspired us along the way.

First, we'd like to thank Icy Liang, who helped lead this project and gave scientific comments and artistic advice along every step of the way.

Next, we'd like to thank all the amazing doctors, including Dr. Steve Jonas, Dr. Francois Luks, and Dr. Yazhen Zhu, who offered their expertise and knowledge to help make this book more scientifically accurate.

Additionally, we'd like to give many thanks to other members of our team, including Patrick Tseng, Gia Clarke, Lynn Zheng, and Audrey Qian, for their advice and suggestions to make the book even better! A special heartfelt thanks goes to Patrick Zhang, whose dedication and hard work on the layout design greatly enhanced the final presentation of this book!

We'd also like to thank our parents for always being enthusiastic and encouraging, providing us all the art materials, along with unconditional love and support.

Finally, we'd like to thank all of you for taking the time to read and enjoy our book!

Clay Anatomy wouldn't be possible without everyone, thank you!

Table of Contents

Hi, I'm Sam, the brain. I will lead you through different parts of the human body and share fun facts along the way.

And I'm Ava, the heart. I will give my heartfelt advice to help you be healthy and happy. Let's go on this journey together!

The Digestive System

Mouth

The mouth is the beginning of the digestive system. As food enters the body through the mouth, your teeth chew the food and your salivary glands produce saliva to help form a bolus. The saliva contains an enzyme that breaks down starches in your food. Afterwards, the food passes through the pharynx, also known as the throat. It carries food, liquid, and air to the esophagus.

Fun fact!

There are about 300 different kinds of bacteria in your mouth. Brushing teeth removes most bacteria and plaque, but not all, which is why flossing and regular dental care are important.

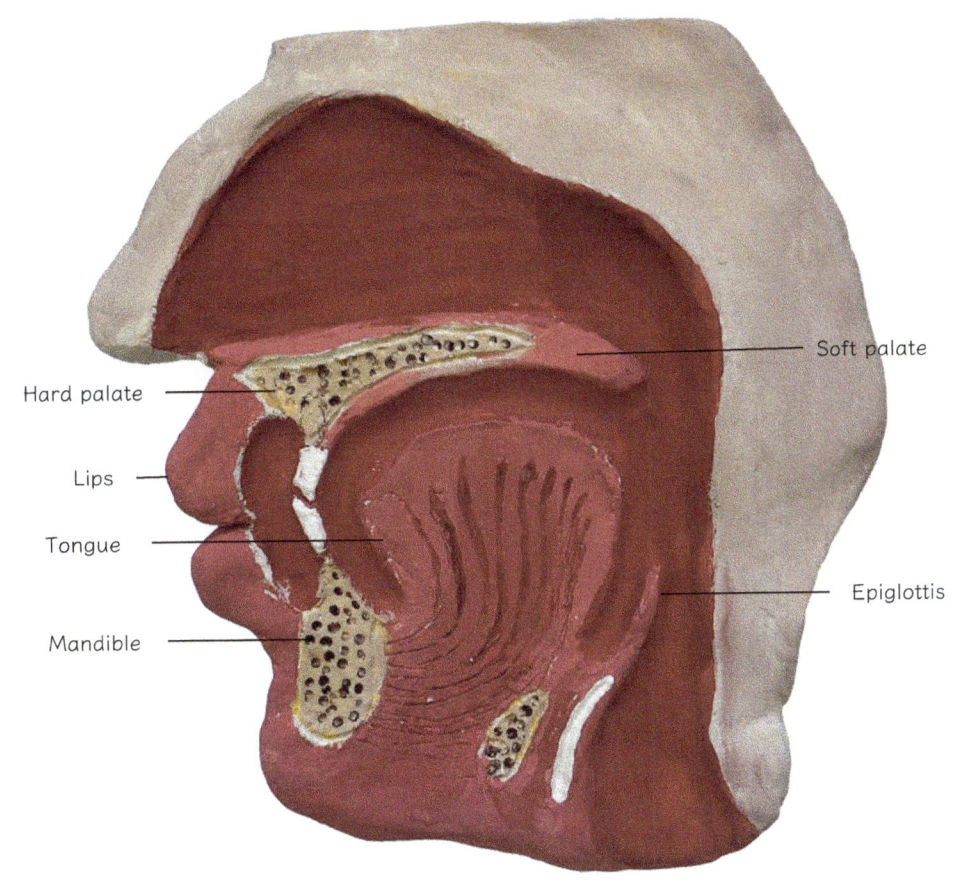

Hard palate

Lips

Tongue

Mandible

Soft palate

Epiglottis

Reminder!

Be mindful of the impact
of your words.

Esophagus

The esophagus carries food from the throat to the stomach. Food doesn't just slide down the esophagus; rather, muscles along the sides squeeze and push the food down.

Upper esophageal sphincter

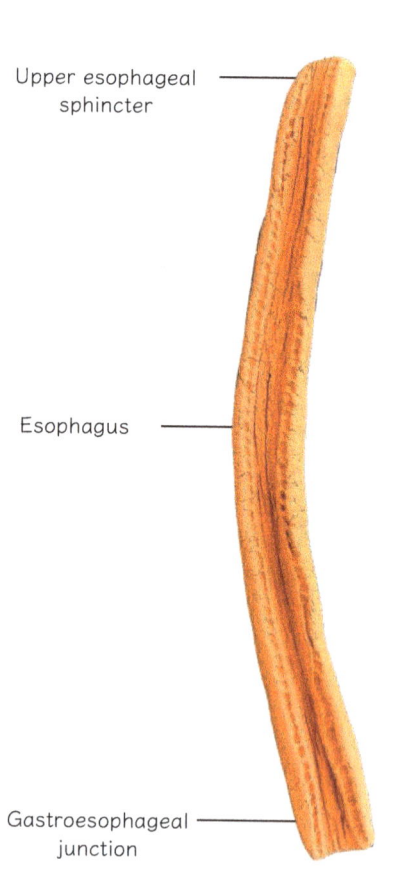

Esophagus

Gastroesophageal junction

Fun Fact!

It takes only seven seconds for the esophagus to propel the food from the throat into the stomach.

Stomach

The stomach digests the food, acting like a blender so we can receive sufficient nutrients to survive. Inside the stomach, gastric acid is produced to denature proteins and activate the enzyme pepsin to begin protein digestion. Stomach acid is so acidic (pH = 2), that a layer of mucus is needed to line the stomach to protect it.

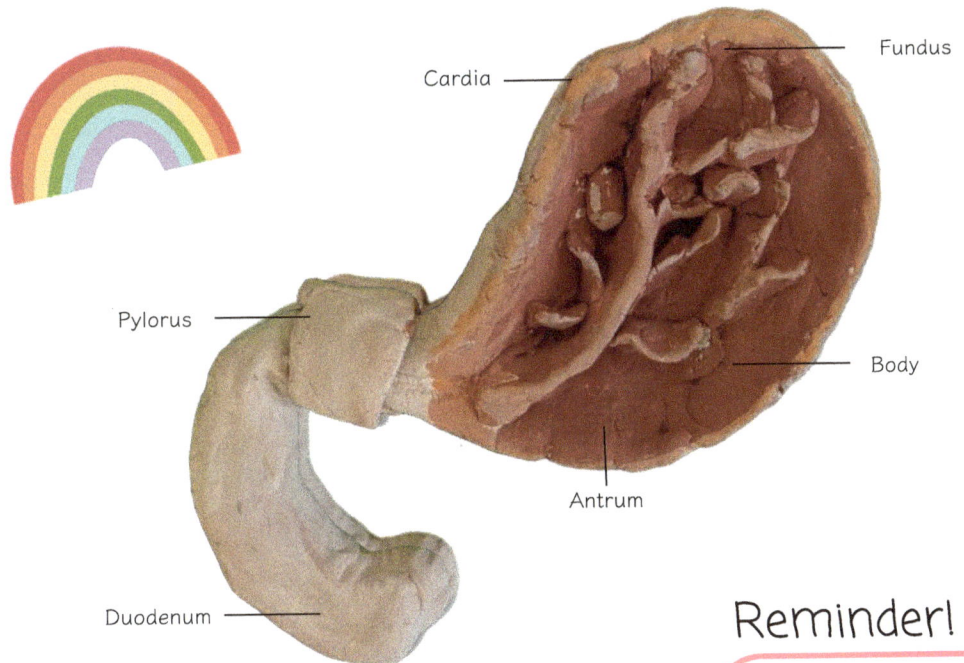

Cardia

Fundus

Pylorus

Body

Antrum

Duodenum

Reminder!

Enjoy a rainbow of foods together: red fruits, yellow grains, and green veggies.

Intestines

The small intestine further digests the food and absorbs nutrients such as vitamins, carbohydrates, minerals, fats, and proteins. These nutrients cross the intestinal lining into the bloodstream and get delivered to cells around the body. The remaining food is pushed through the large intestine where the remaining water and salt is sucked out. The waste that is left behind is eliminated through the rectum.

Fun Fact!

The small intestine is three times the length of an average adult.

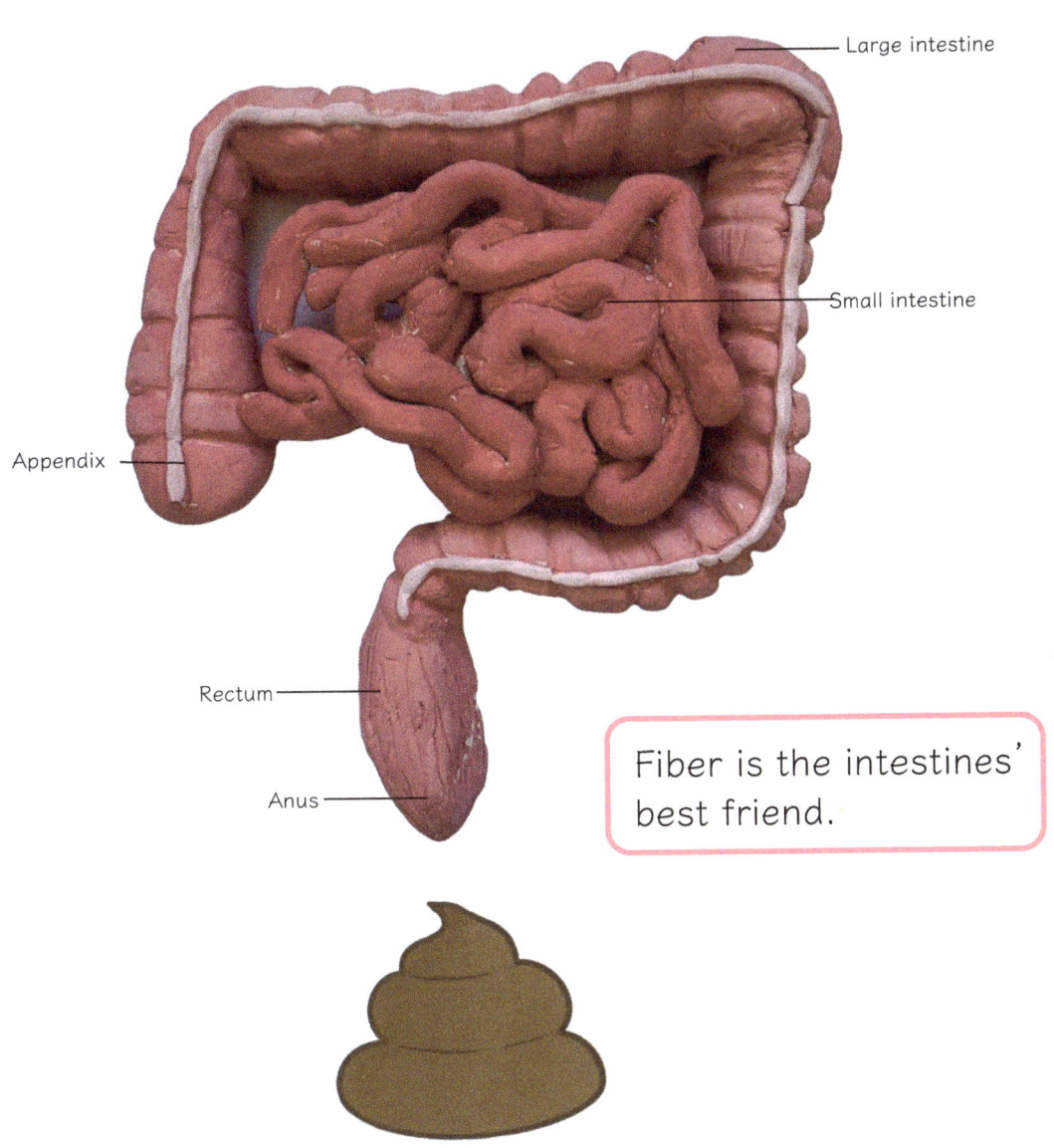

Large intestine

Small intestine

Appendix

Rectum

Anus

Fiber is the intestines' best friend.

Liver + Gallbladder

Front View

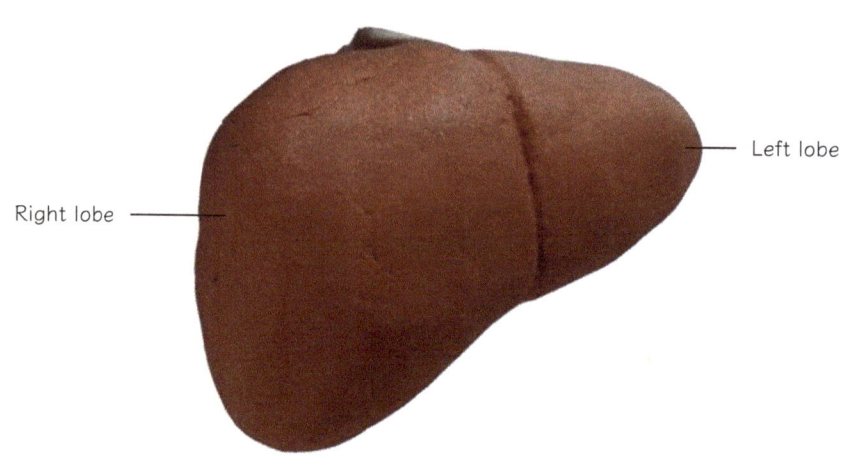

Right lobe —

— Left lobe

The liver acts as the main purifier for blood by detoxifying chemicals and drugs and metabolizing waste. Much of the blood from the digestive tract passes through the liver, where cells in the sinusoid channels called Kupffer cells detoxify the blood by filtering it. The liver also helps with digestion and stores important glucose, vitamins, and minerals. The gallbladder, located beside the liver, stores bile produced by the liver and releases it into the first section of the small intestine, where it breaks down fats during digestion.

Back View

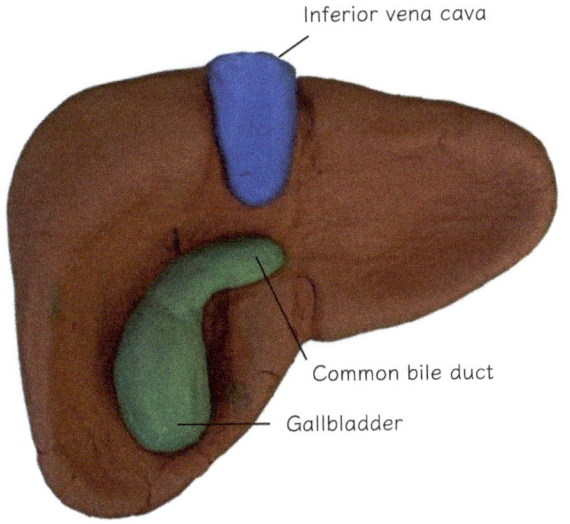

Inferior vena cava

Common bile duct

Gallbladder

Reminder!

I dislike alcohol and fatty foods.

Liver Cirrhosis

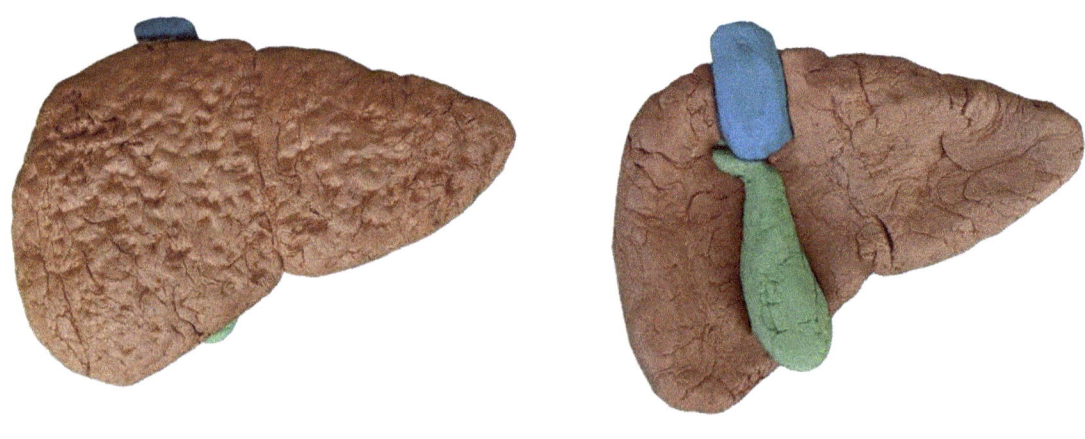

When the liver is continuously filtering alcohol, being subjected to a virus such as HBV, or other long-term destruction, the liver cells become damaged and die off. The regenerated liver cells and their stromal cells form regenerative nodules, causing the bumpy texture. Heaps of these dead cells form scar tissue, also known as cirrhosis. This process is permanent and irreversible.

Liver Cancer

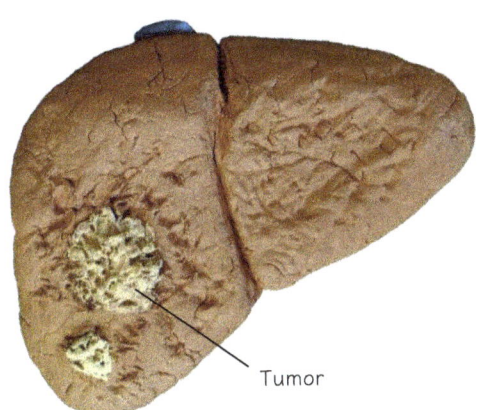

Tumor

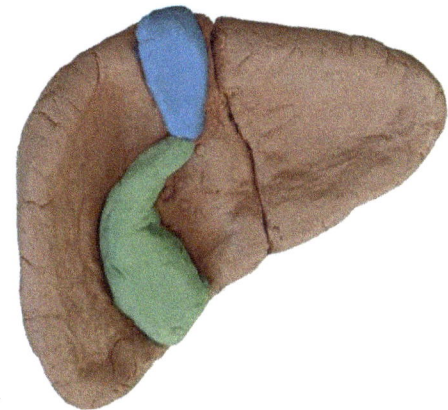

Liver cancer is the third leading cause of cancer-related deaths worldwide. It is associated with fatty liver disease, alcoholism, and hepatitis B or C. The cancer cells grow rapidly and form tumors that can spread throughout the liver and migrate or metastasize to other organs.

Pancreas

Reminder!

Trick-or-treat,
eat less sweets.

The pancreas helps regulate blood sugar by releasing insulin and glucagon. It also helps with digestion by secreting a variety of enzymes and hormones that are released into the small intestine. For example, insulin and glucagon are made in the pancreas and released to help regulate blood glucose (sugar) levels. If the pancreas doesn't produce enough insulin or respond well to it, the person may have diabetes.

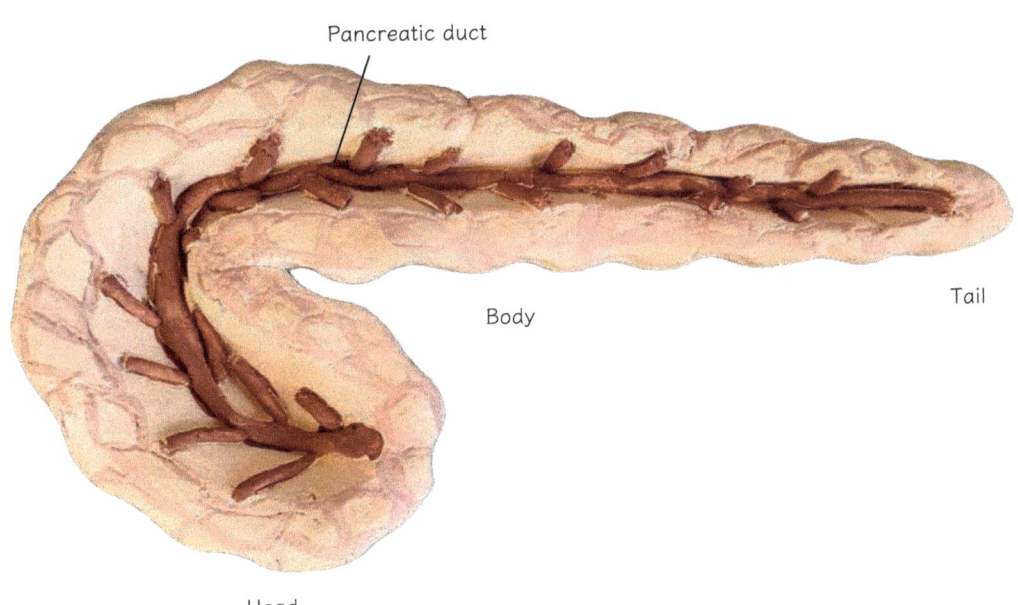

Pancreatic duct

Body

Tail

Head

Reminder!

Remember to follow your heart.

The Cardiovascular System

Heart

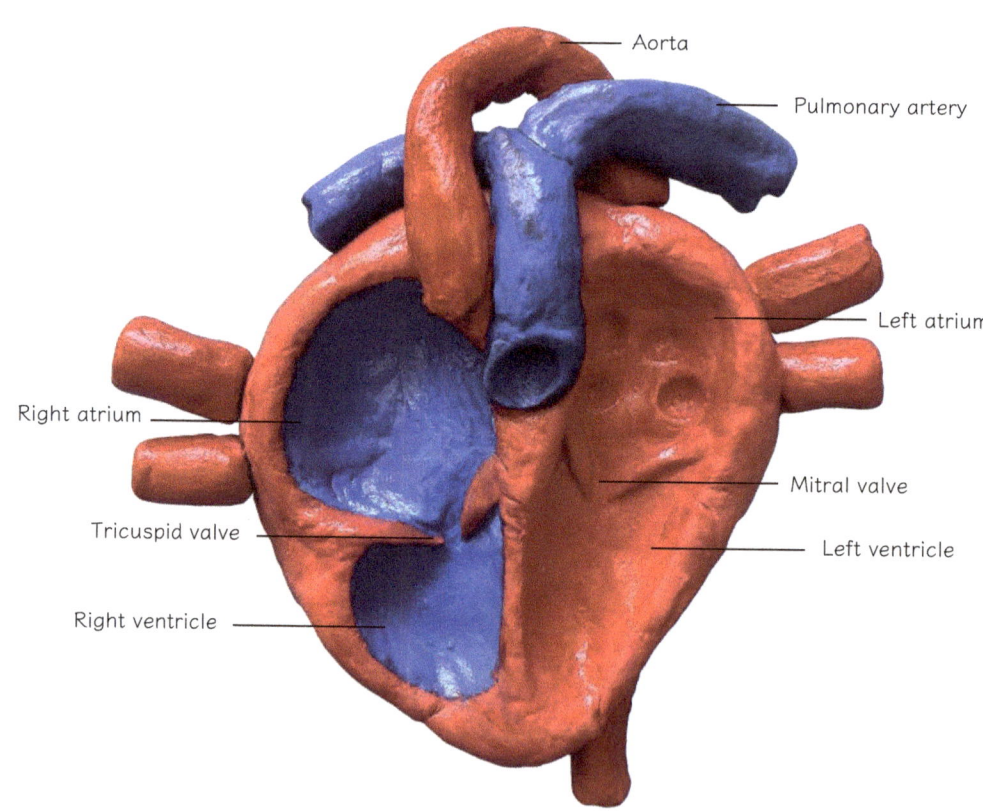

Aorta

Pulmonary artery

Left atrium

Right atrium

Mitral valve

Tricuspid valve

Left ventricle

Right ventricle

And it pumps 2,000 gallons of blood— enough to fill two swimming pools!

The main function of the heart is to pump oxygenated blood to parts of your body and to deliver waste products to the lungs for removal. The organ is made up of four chambers. The right atrium receives blood and pumps the blood into the right ventricle, where it is pumped into the lungs and loaded with oxygen. The oxygenated blood is then delivered into the left atrium and afterward the left ventricle. The cycle continues as the left ventricle pumps the blood to different parts of your body.

Blood Vessels

Blood vessels act as the body's transportation system for blood. Oxygen-rich blood is pumped from the heart to arteries and then smaller blood vessels known as capillaries. The capillaries deliver oxygen and nutrients to various organs in the body. Veins then carry most of the deoxygenated blood (rich in carbon dioxide) back to the heart, which pumps it to the lungs for removal of carbon dioxide and uptake of oxygen.

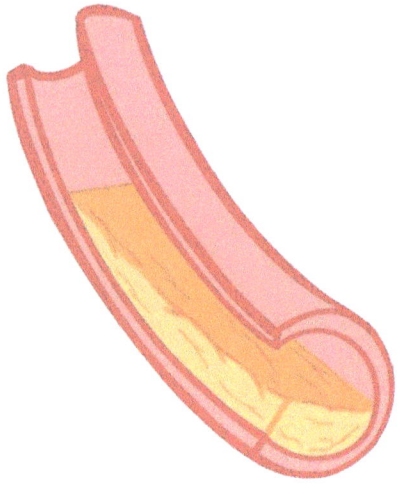

Fun Fact!

Too much cholesterol is unhealthy because it causes plaque to build up inside blood vessels. When the plaque ruptures, a blood clot can form and block the artery. This may cause a heart attack.

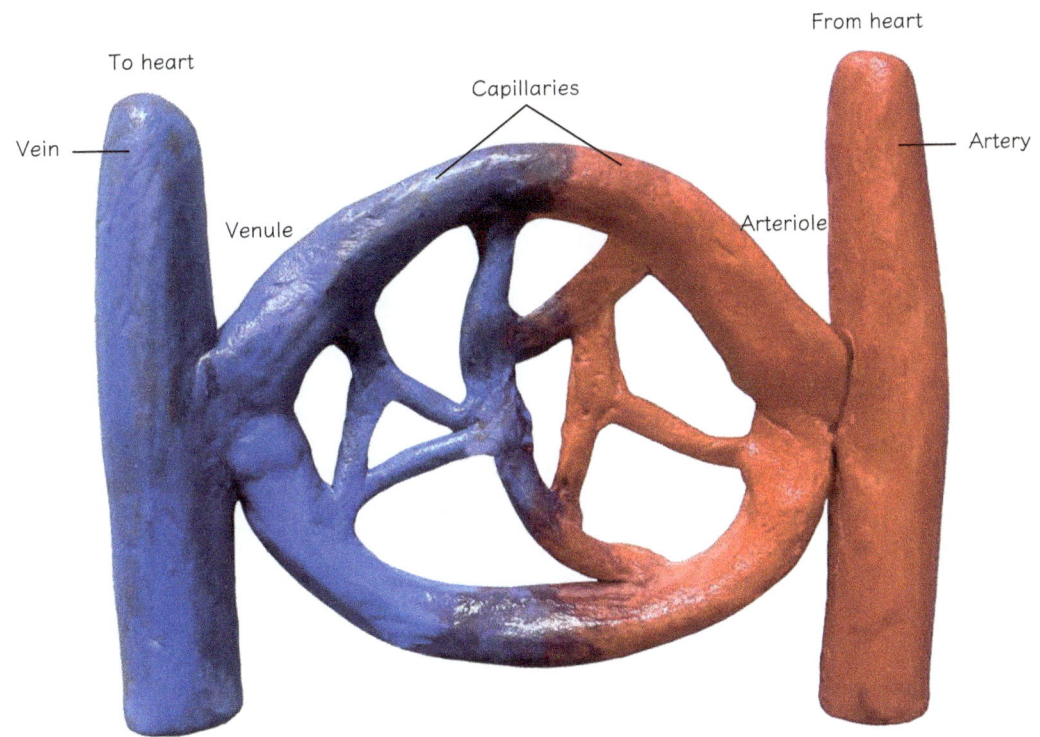

To heart

From heart

Vein

Artery

Capillaries

Venule

Arteriole

Reminder!

Exchange your ideas with one another.

Reminder!

Take a deep breath and enjoy the moment!

The Respiratory System

21

Nose

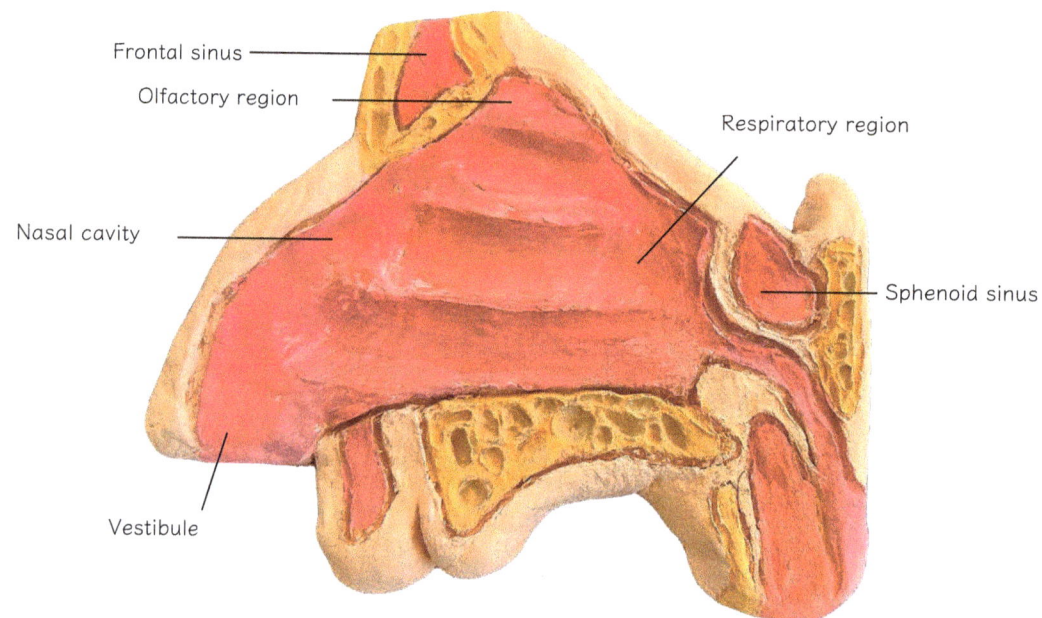

Frontal sinus

Olfactory region

Respiratory region

Nasal cavity

Sphenoid sinus

Vestibule

The nose is the main gateway to your respiratory system. It filters, warms, and moistens the air, getting it ready for your throat and lungs. Not only does the nose allow you to smell, but it also produces mucus which traps pathogens before they enter the lungs.

Fun Fact!

In adults, about 18,000 to 20,000 liters of air pass through the nose on average daily.

Reminder!

Take a deep breath and enjoy the moment!

Lungs

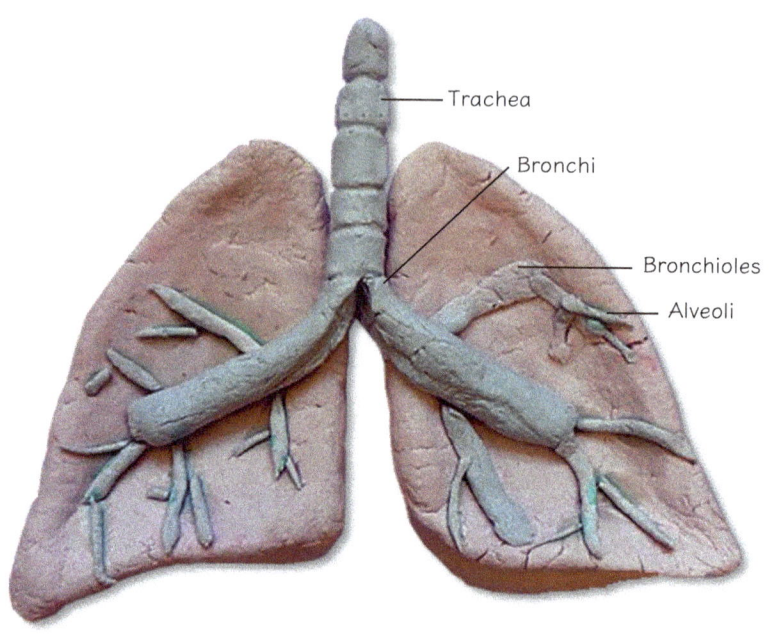

The lungs help the body breathe by delivering oxygen and removing carbon dioxide from the body. The lungs are located on either side of your heart inside your chest cavities. Each lung is divided into lobes; while the right lung is divided into three lobes, the left lung is only divided into two lobes.

Smoking Lung

Reminder!

Don't smoke, and stay away from secondhand smoke.

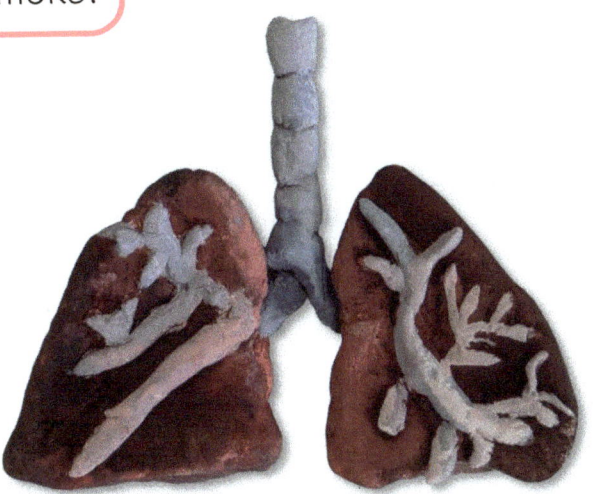

Healthy lungs are light pink, while a smoker's lungs are darker colored. Smoke contains carbon monoxide, a deadly gas that replaces oxygen in the blood. It irritates the lungs and causes them to work less efficiently. A person that smokes may feel out of breath more easily after physical activity.

Lung Cancer

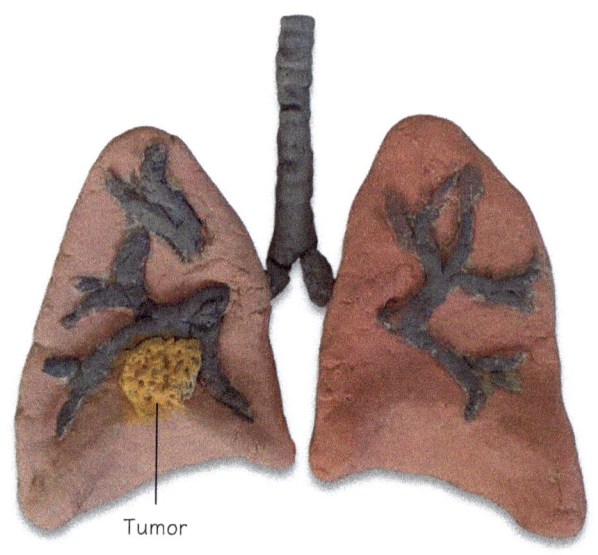

Tumor

Lung cancer is a type of cancer that resides in the lungs. Cancer is caused when abnormal cells multiply uncontrollably, forming tumors. These tumors can interfere with daily activities, causing the person to have shortness of breath, show a persistent cough, or feel hoarse and tired.

The Nervous System

Brain

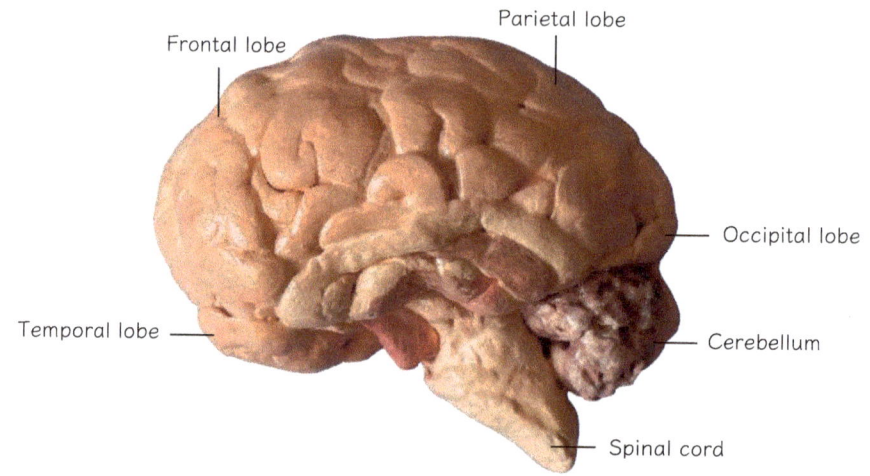

Frontal lobe

Parietal lobe

Occipital lobe

Temporal lobe

Cerebellum

Spinal cord

The brain is located within the protective covering of the skull. It controls everything you do from talking and thinking to feeling emotions and learning. The brain consists of four lobes. The frontal lobe is involved in conveying personality and making decisions. Next to it is the parietal lobe, which interprets pain and touch. The occipital lobe (back of the brain) controls vision, and the temporal lobes (sides of the brain) control short-term memory and musical rhythm.

In your brain, there are cells called neurons that work like messengers. They relay information using electrical signals. When you learn a new task such as riding a bike or swimming, the neurons make new connections with each other. As you practice the skill more and more, these pathways grow stronger and messages travel more easily through the brain.

Reminder!

Nuts, fish, berries, and greens are "brain foods" and can improve your concentration and memory.

Spinal cord

The spinal cord, the supporting structure of the human torso, plays an important role in your everyday life. The brain sends messages down the spinal cord and then to different areas of the body. The spinal cord also regulates reflexes. When someone touches a hot pot, they will pull away immediately. That's because the spinal cord acts instinctively to keep you safe.

Fun Fact!

The spinal cord reaches its full length in early childhood, but it continues to develop connections as you grow.

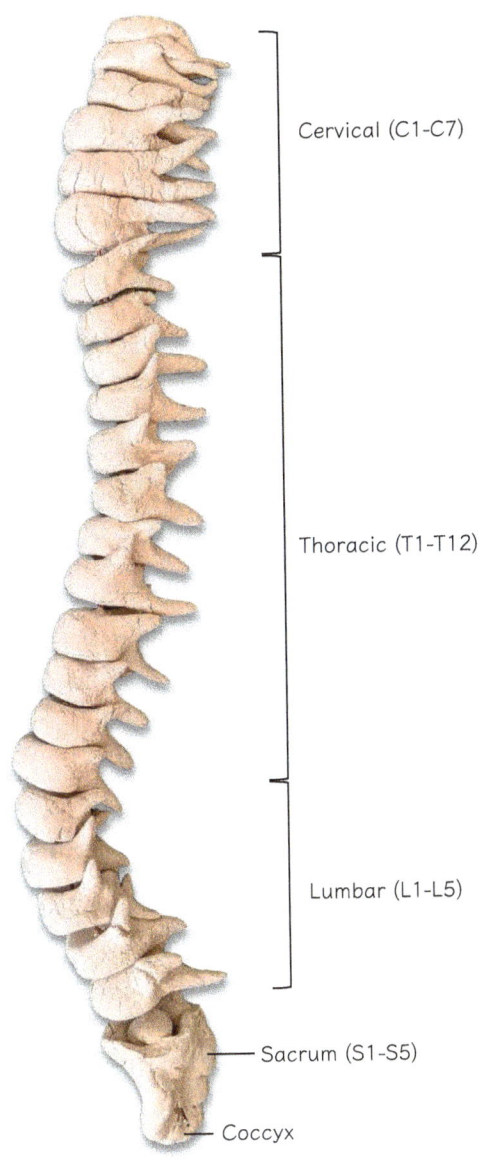

Cervical (C1-C7)

Thoracic (T1-T12)

Lumbar (L1-L5)

Sacrum (S1-S5)

Coccyx

Reminder!

Alone each bone is strong; together, they are stronger.

Eye

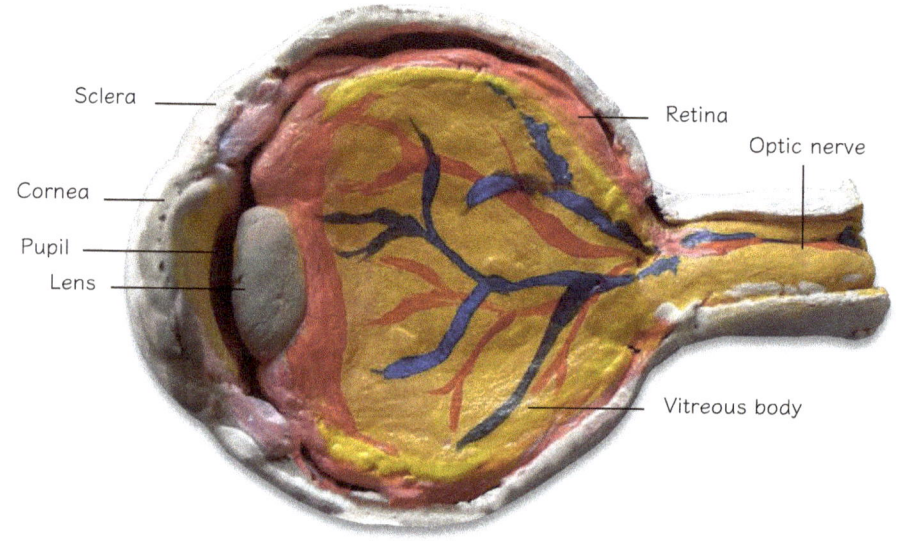

Sclera

Cornea

Pupil

Lens

Retina

Optic nerve

Vitreous body

The eye is a sensory organ that helps us see the world. Light enters the eye through the pupil and gets converted into a nerve impulse. The optic nerve, located at the back of the eye, then sends these signals to the brain which processes what we see.

The cells that process color are called cone cells, and humans have three kinds of cones: red, blue, and green. The mantis shrimp has sixteen types of cones, allowing them to see many more colors than us. While cones are active in the light, other cells called rods are active in the dark. They help us see in low light levels.

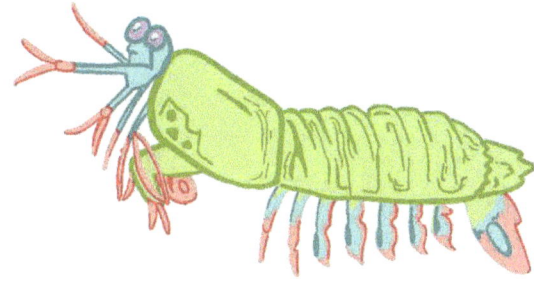

Mantis shrimp

Reminder!

Look for the beauty in everything you see.

Reminder!

Stay calm; you don't need to be nervous.

The Urinary System

Kidney

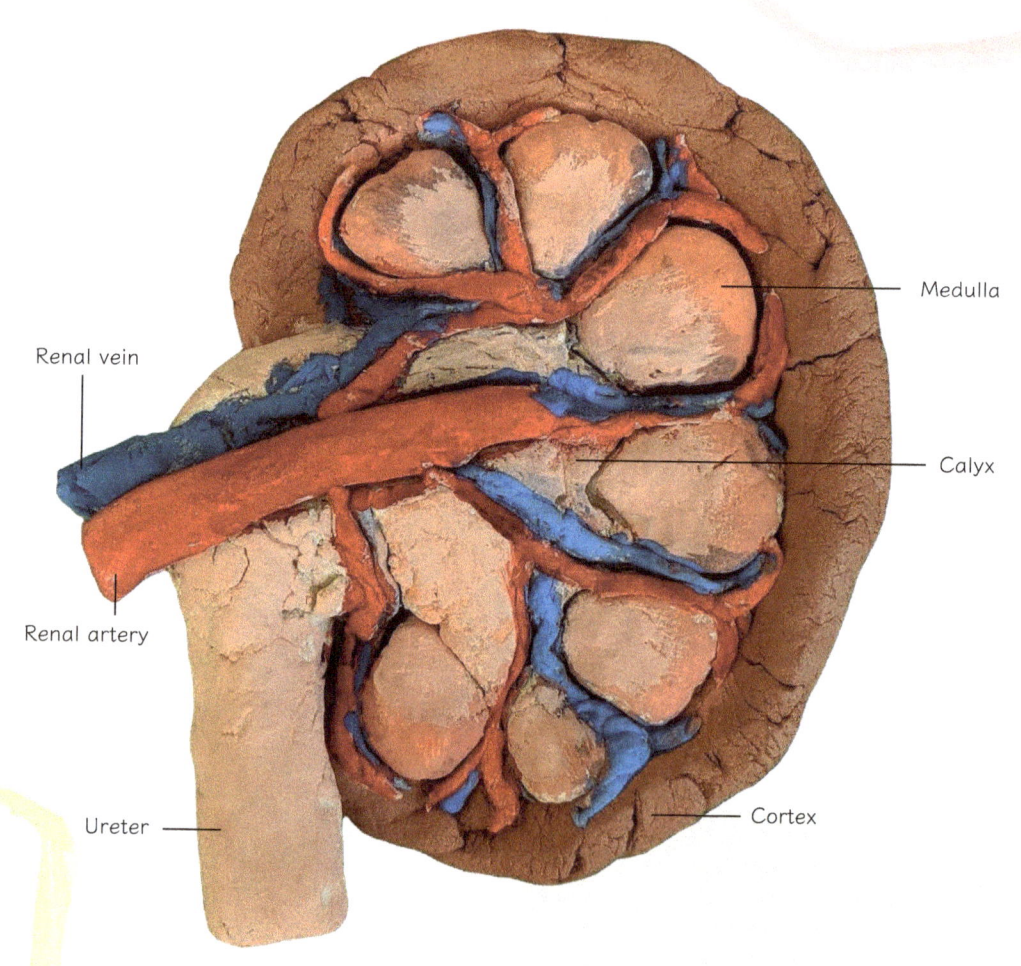

Renal vein

Renal artery

Ureter

Medulla

Calyx

Cortex

Reminder!

Drink lots of water.

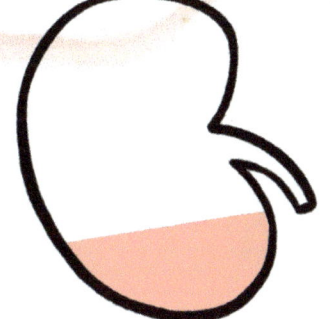

The kidneys receive about 20% of the body's blood flow at any given time!

The two kidneys function like washing machines for the blood. When blood flows through the renal artery to the renal medulla, small nephrons filter the blood to keep useful minerals and to remove extra water and waste. The waste (urine) then exits the kidney through the ureter and goes to the bladder. The kidneys play an important role to maintain balanced water and salt levels.

Fun Fact!

The bladder can show whether you're drinking enough water, turning darker or lighter yellow as an indicator.

Bladder

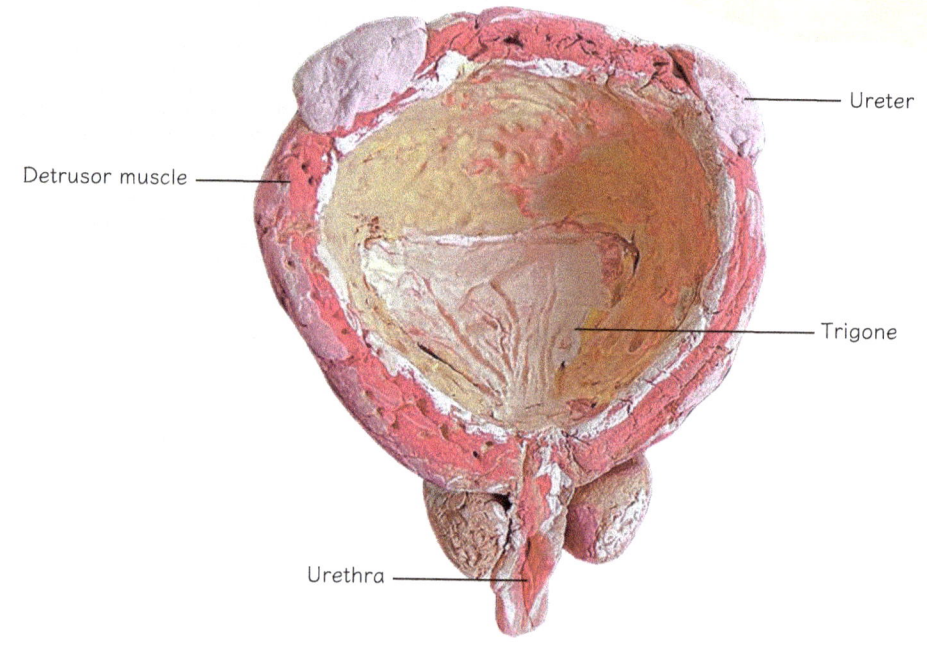

Ureter

Detrusor muscle

Trigone

Urethra

The bladder is a round organ where urine is collected for excretion. After we consume food, the useful bacteria and nutrients from the food are absorbed by the body, while the waste products are removed. Urine disposes of excess ions and compounds such as chloride, sodium, potassium, sulfate, ammonium, and phosphate.

The Immune System

Bone Marrow

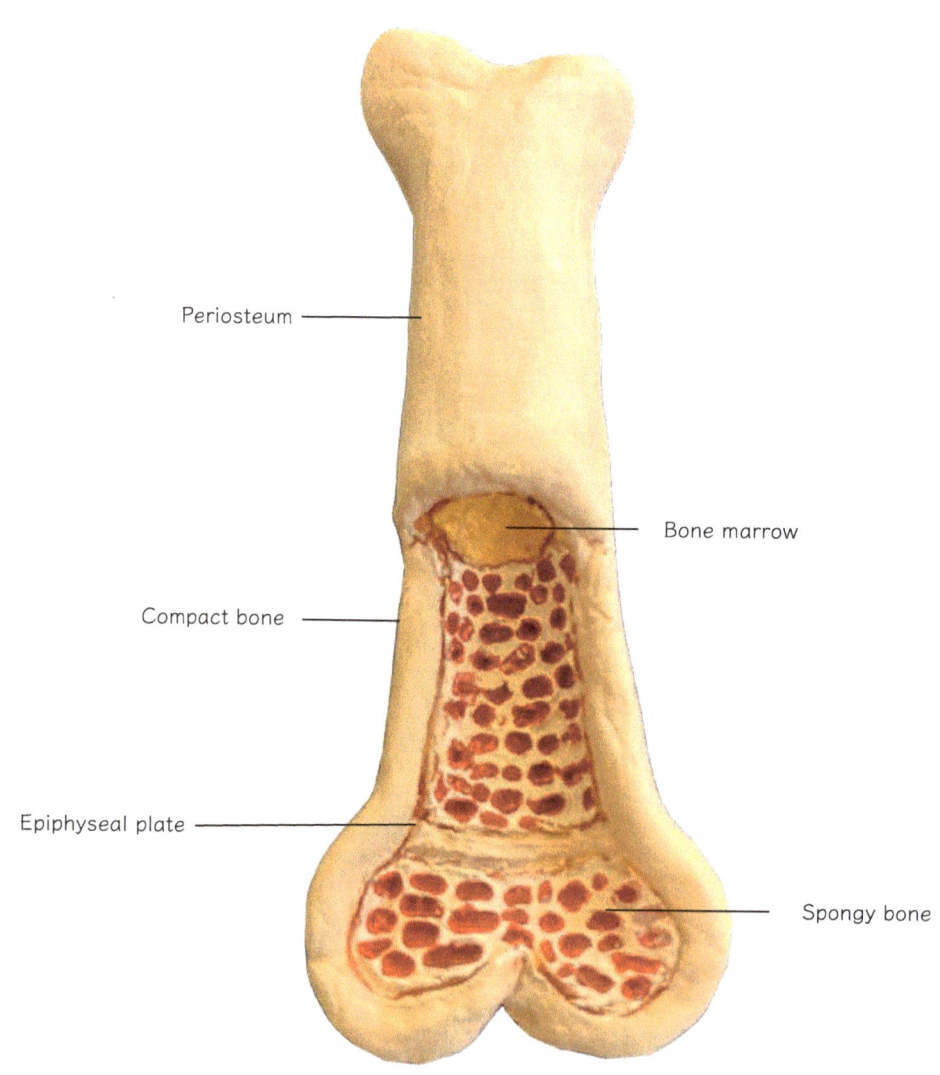

Periosteum

Bone marrow

Compact bone

Epiphyseal plate

Spongy bone

The bone marrow produces hematopoietic stem cells, which can turn into any cell type found in the blood. They can become red blood cells to deliver oxygen to other tissues, or white blood cells to destroy infections. Platelets are formed from large bone marrow cells called megakaryocytes and enter the bloodstream to help stop bleeding.

Types of Blood Cells

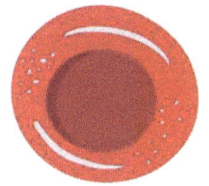

Red blood cell

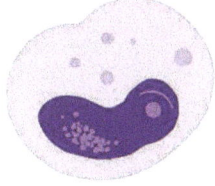

Monocyte

Leukemia is a cancer of the blood cells. It affects the body's white blood cells, making it harder for the immune system to control infections.

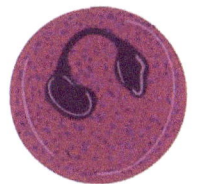

Eosinophil

Reminder!

Don't fear getting vaccines. They boost your immune system.

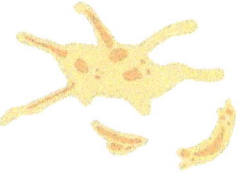

Platelet

Spleen

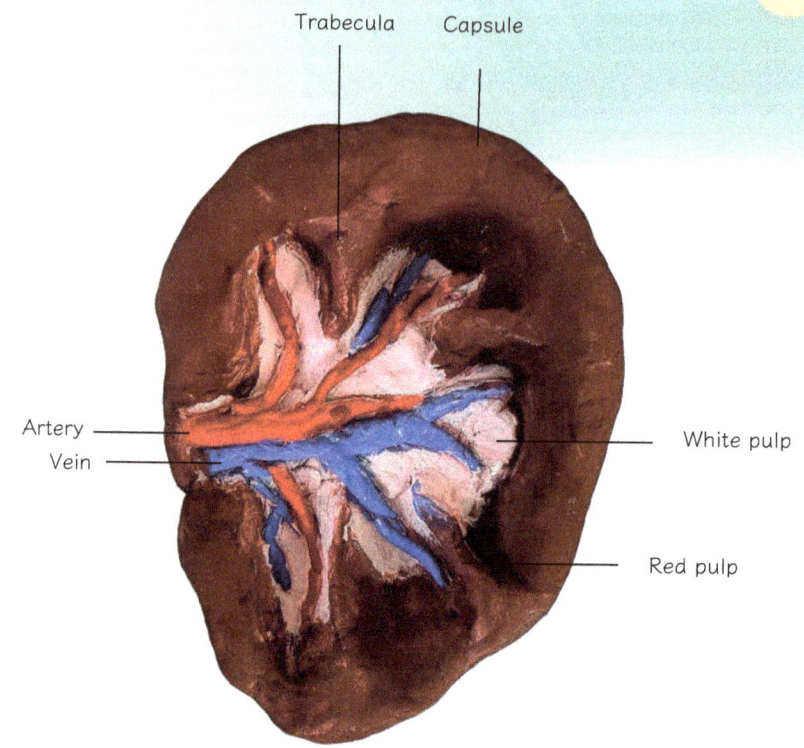

Trabecula   Capsule

Artery

Vein

White pulp

Red pulp

The spleen stores blood and filters it as blood circulates throughout the human body. The two main parts of the spleen are the white pulp and red pulp. White pulp produces white blood cells such as lymphocytes and macrophages, which fight off infections. Red pulp filters out old and damaged blood cells. The spleen helps you stay healthy by destroying bacteria and viruses that enter your body.

Lymph Node

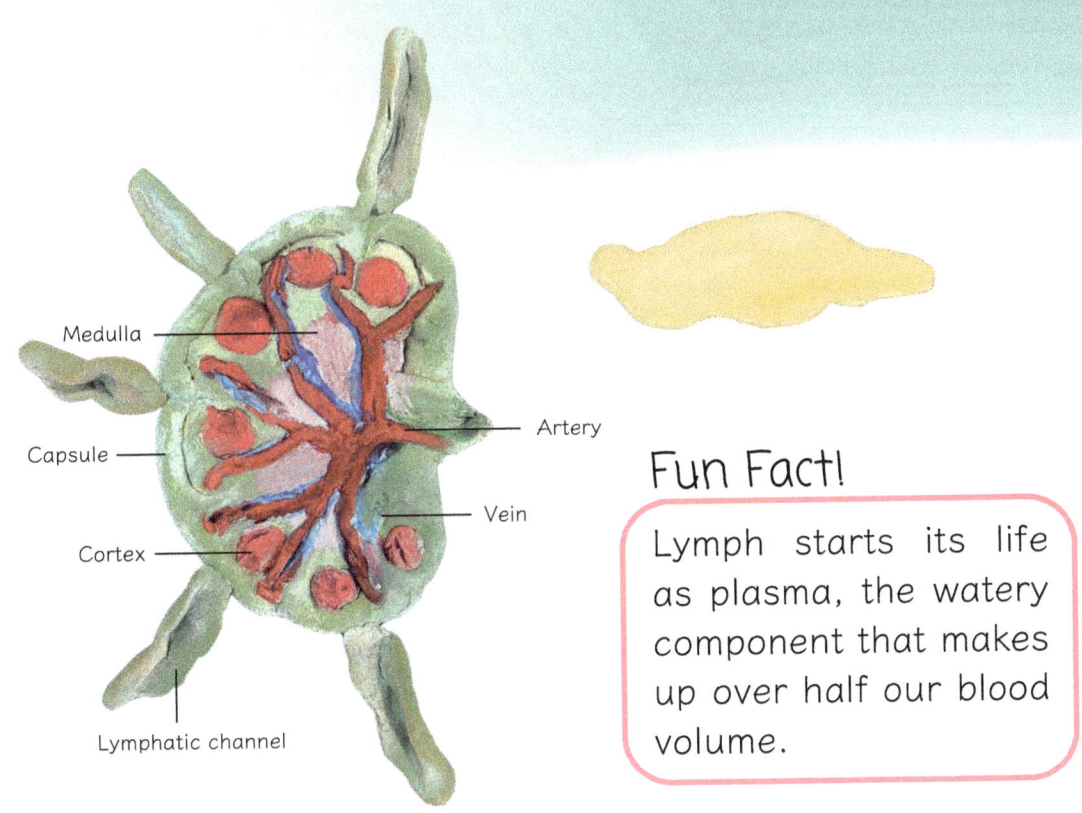

Medulla

Capsule

Cortex

Lymphatic channel

Artery

Vein

Fun Fact!

Lymph starts its life as plasma, the watery component that makes up over half our blood volume.

These little lymph nodes are the size and shape of a pea, and they're found throughout the body. They hold white blood cells and filter lymph, a clear-to-white fluid. Damaged cells and cancer cells are removed from the lymphatic fluid. Lymph nodes become swollen when the body has an infection or an illness. It's their natural reaction!

Thymus

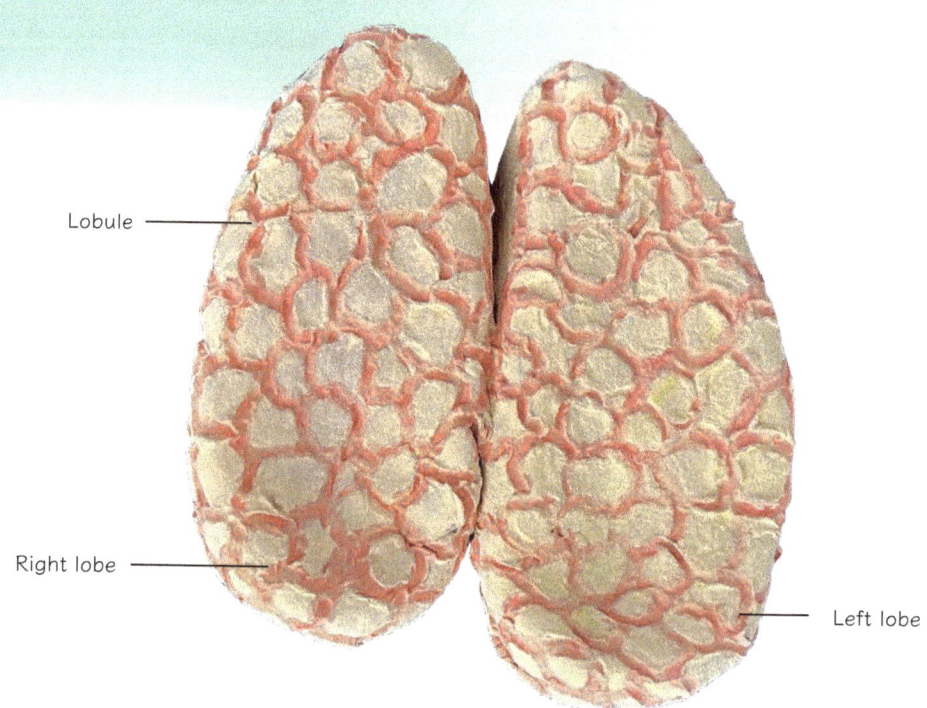

Lobule

Right lobe

Left lobe

The thymus produces white blood cells, specifically T cells. T cells are aptly named because they're made in the thymus. These cells recognize and fight against germs and diseases that may make your body sick. The thymus is most active when you are a child and produces all the T cells you need by the time you reach puberty. Afterward, the thymus slowly shrinks and is replaced by fat. By the time you are 75 years old, it will have reduced to fatty tissue.

Skin

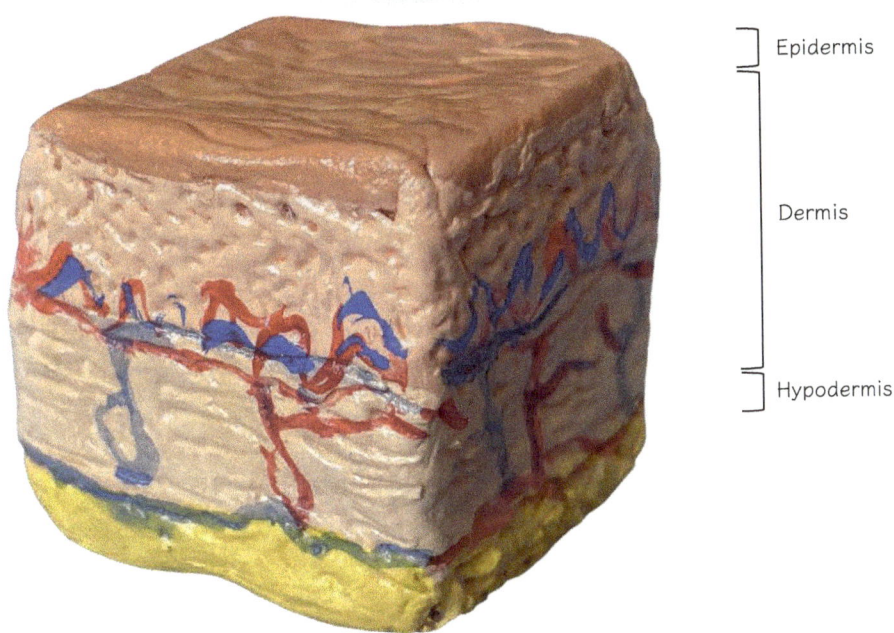

Epidermis

Dermis

Hypodermis

Fun Fact!

The skin is the largest organ.

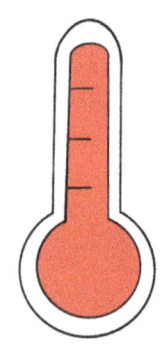

Skin is much more than just the outermost layer of the human body. It acts as a sensory organ involving common senses such as touch and detection of temperature. It also acts as a protective barrier for the body, shielding against physical injuries, as well as mechanical, thermal, and other dangerous substances. The thickest part of your skin is on the bottom of the feet, and the thinnest part is your eyelid. Your skin also renews itself every 28 days.

Reminder!

Stay in the shade, no more UV rays.

The Endocrine System

Thyroid

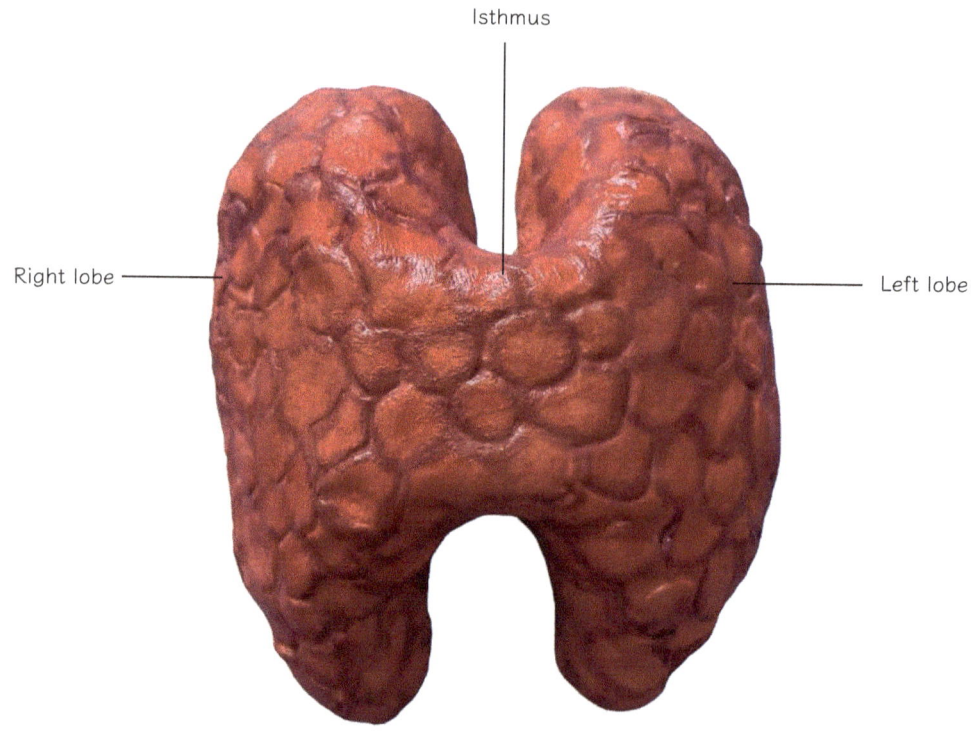

Isthmus

Right lobe

Left lobe

Reminder!

Find balance in
your life.

Located in front of the throat, the thyroid gland produces hormones that regulate the growth, metabolism, and development of the body. These hormones are like messengers that keep your body in balance. It tells your body how much energy to use and how fast to digest your food. It also raises and lowers your body temperature, as well as slows down or speeds up your heart rate.

Fun Fact!

A healthy supply of iodine from a good diet is essential to the thyroid's functions.

Ovaries

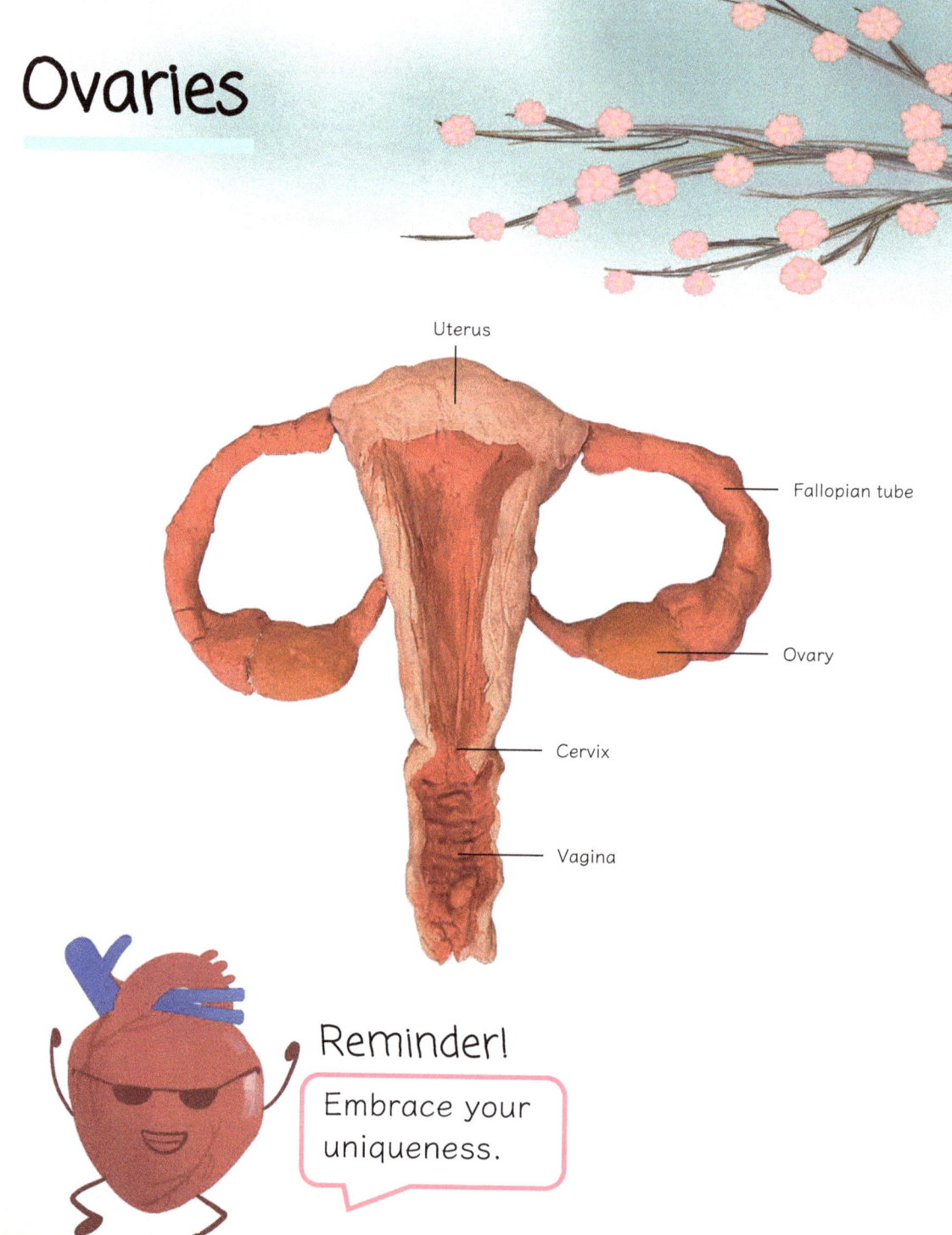

Uterus

Fallopian tube

Ovary

Cervix

Vagina

Reminder!

Embrace your uniqueness.

Fun Fact!

A baby is born with all the eggs she will have for the rest of her life.

The ovaries produce and store eggs, as well as make hormones such as estrogen. Estrogen regulates the menstrual cycle and affects the development of secondary sex characteristics. Some of its key roles include pregnancy, fertility, and menopause.

Testes

The testes' main functions are to produce and store sperm. They're also important for producing male hormones such as testosterone and androgens. The prostate's main function is to produce a liquid that nourishes and transports sperm through seminal fluid.

Fun Fact!

Sharks have two claspers, which serve the same purpose as penises: for internal fertilization.

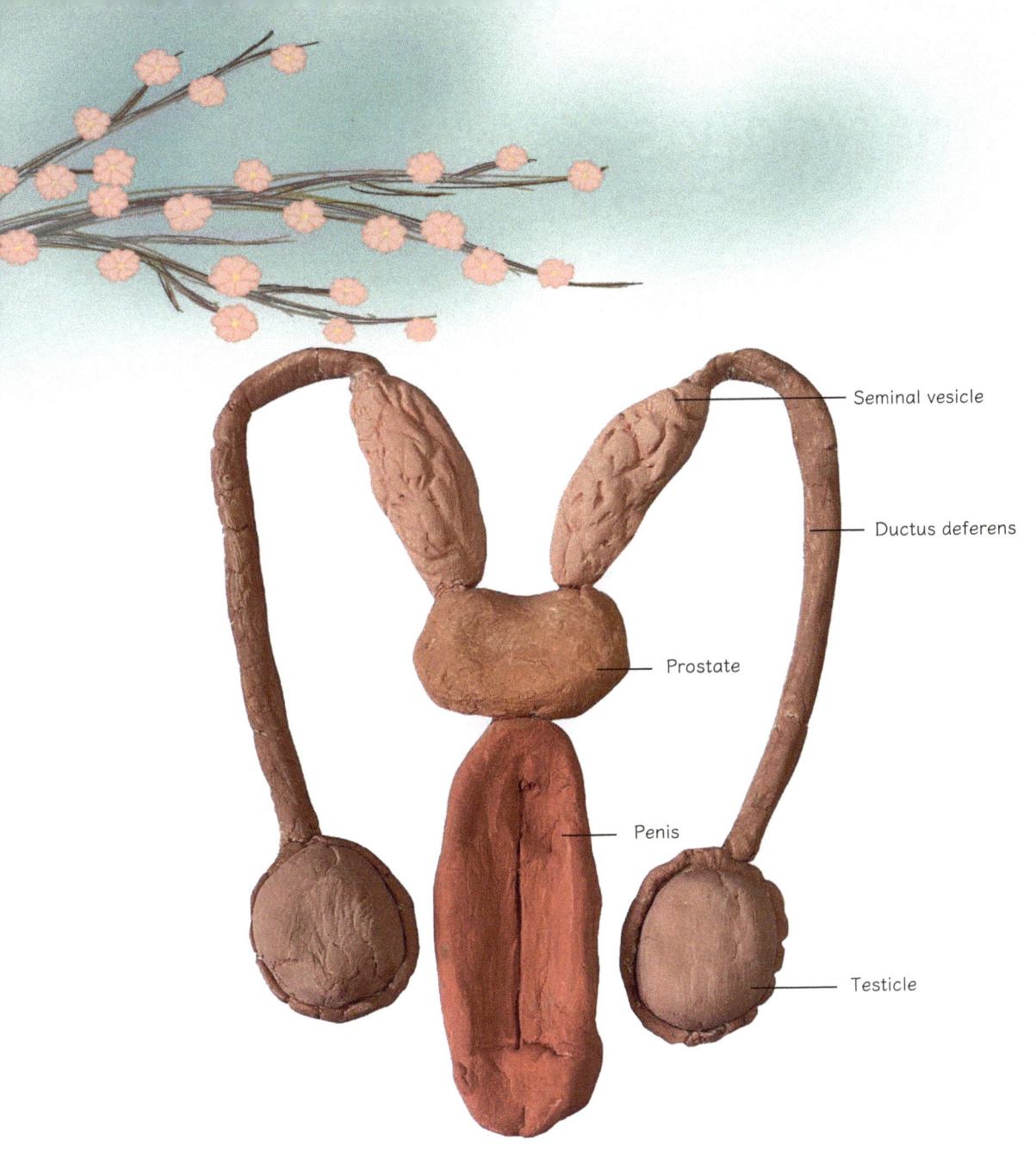

Seminal vesicle

Ductus deferens

Prostate

Penis

Testicle

Other glands

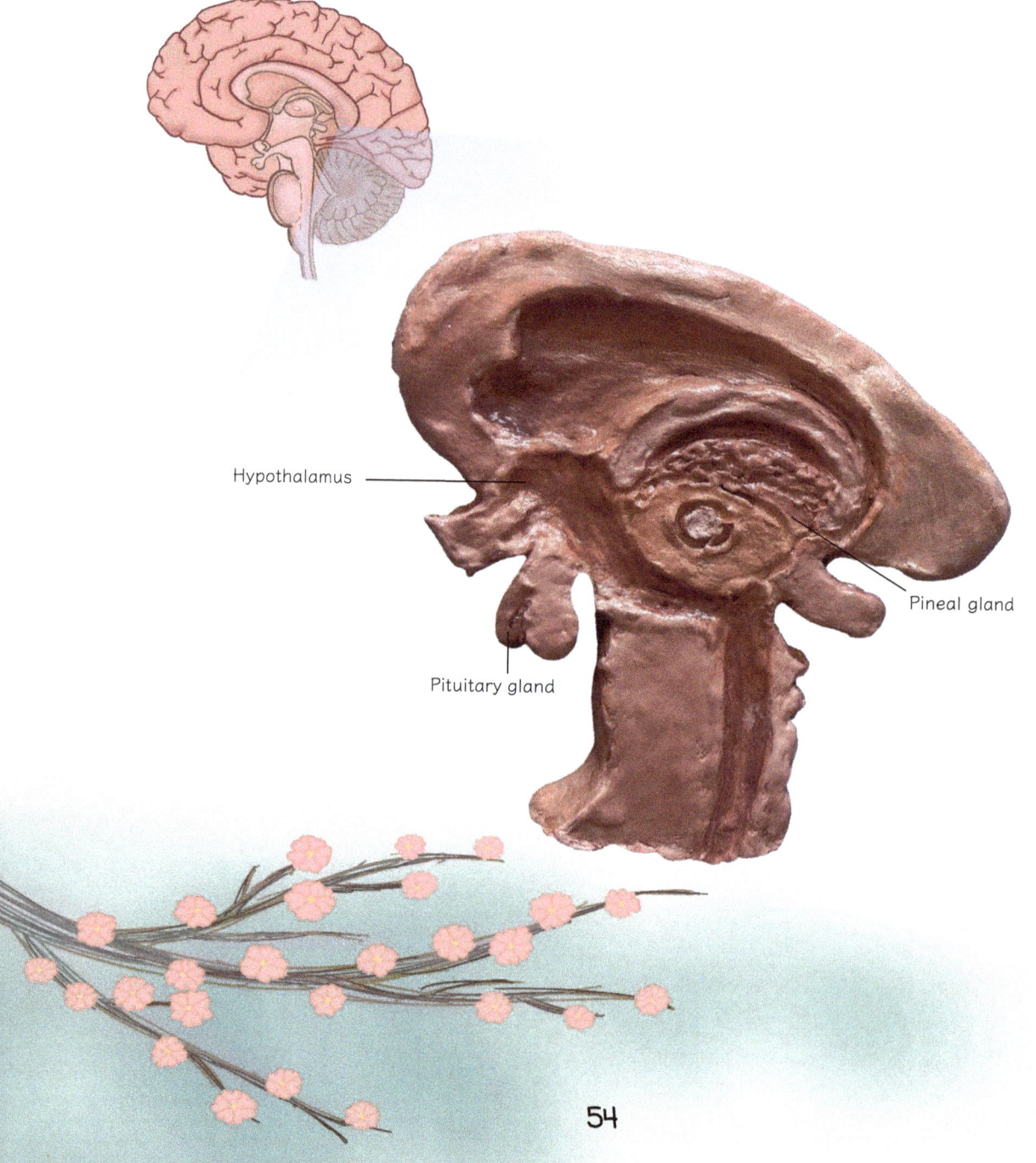

Hypothalamus

Pineal gland

Pituitary gland

The hypothalamus regulates your body temperature, hunger, and mood, as well as many other functions. Some hormones are released from the hypothalamus to control the pituitary gland.

The pituitary gland is divided into two main lobes known as the anterior and posterior regions. It releas-es many hormones that can act on other glands such as the thyroid, adrenal gland, and ovaries/testes.

The pineal gland is a very small gland that controls your circadian rhythm or your internal body clock. It tells you when to sleep and wake up daily.

Adrenal gland

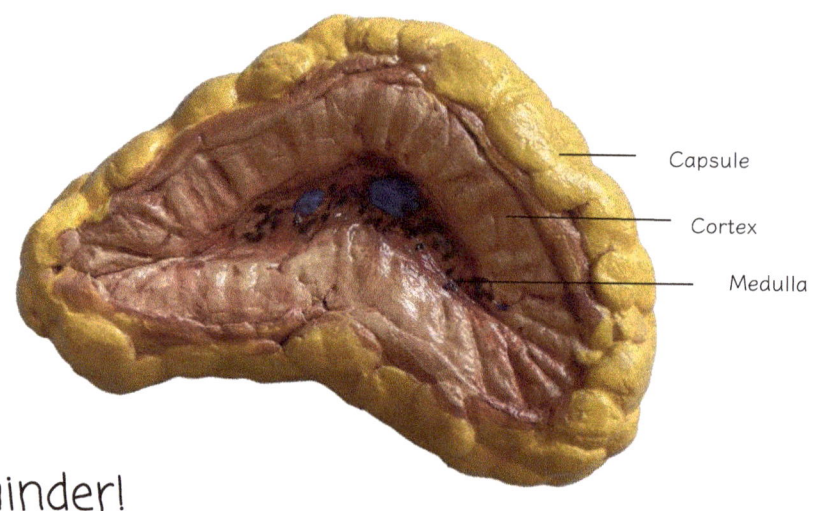

Capsule

Cortex

Medulla

Reminder!

Be brave and face
your challenges!

The adrenal gland is located above each of the two kidneys and produces hormones such as aldosterone, cortisol, and adrenaline. Hormone levels can shift to keep your body in balance. When encountering a scary or stressful situation, the body releases adrenaline to drive a fight or flight response. For example, when encountering a bear, one may fight back or run away on the spur of the moment. In this situation, the adrenal gland gives you an energy boost and increases your blood pressure and heart rate.

The Musculoskeletal System

Bone & Muscle

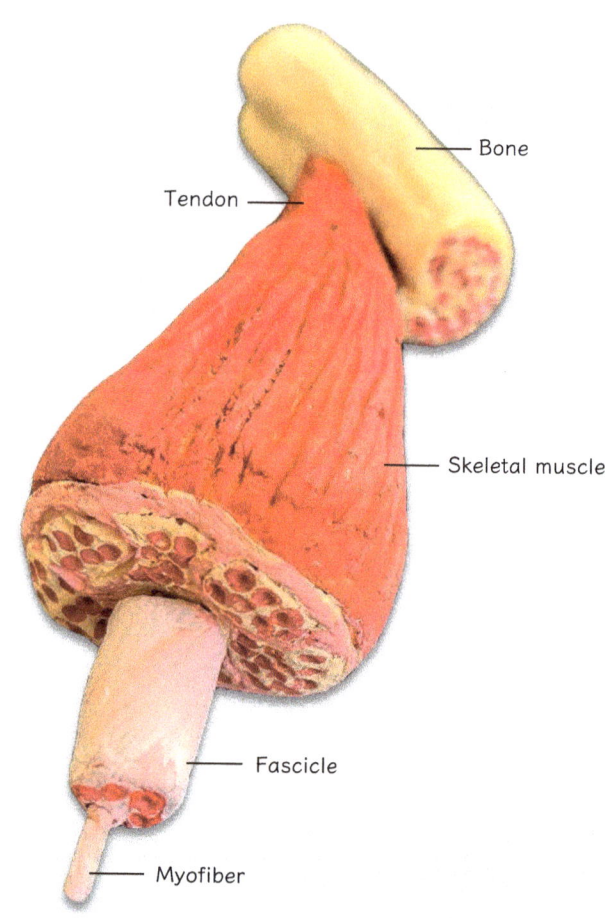

Bone

Tendon

Skeletal muscle

Fascicle

Myofiber

Fun Fact!

A person is born with about 300 bones, but over time, they fuse together to form around 206 bones.

Reminder!

Exercise and build your muscles.

Fun Fact!

The smallest muscle is the stapedius, and the smallest bone is the stapes. Both are located in the ear.

Bones and muscles are two parts of the musculoskeletal system. Bones and cartilage support the body and protect internal organs. Bones also store about 99% of the body's calcium, and they grow stronger when you drink milk. Muscles help a person move around and control heartbeat, digestion, vision, and breathing.

The Placenta:
A Temporary Organ

Placenta Accreta

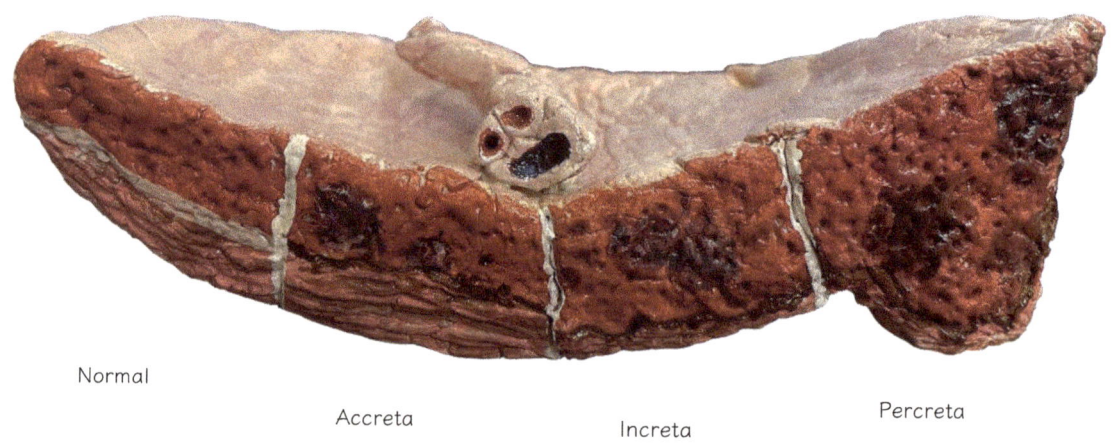

Normal

Accreta

Increta

Percreta

The placenta develops in the uterus during pregnancy. Normally, after childbirth, the placenta separates from the uterine wall, but with placenta accreta, it remains attached. This may cause a risk of future complications related to childbirth.

About Us

About the Authors:

Hello there! I'm Lucy, an artist who enjoys art and writing. Writing this book has been an unforgettable adventure. Not only did I dive deep into the artistic process and learn about the amazing world of human anatomy, but I also embraced the joys of collaboration and the challenges of managing my schedule after school. My favorite organ is the bone marrow because it's so versatile. Combining clay with science was an enjoyable new experience for me, and I've grown to appreciate how art makes science fun and easy to understand.

I'm Angela, and I enjoy artistic swimming, science, and art. I have enjoyed working with clay in the past, so being able to work with clay again was extremely exciting. My favorite part of this book was making the clay model of the mouth because it had many different details. Throughout this experience, I have gained a lot of knowledge about human anatomy, working and painting with clay, and teamwork.

About the Editor:

I am Icy and I helped edit the book. My interests lie within the intersection of biology and art, so I had lots of fun planning and guiding everyone to create the book. I hope this book can reach the hands of many people and spread knowledge in a fun and understandable way.

How We Did It

First, we chose an organ in the body and did some initial research on its appearance and functions. We would look at images and models of the chosen organ to become familiar with its different components and appearance. Then, we used clay to form and shape the clay organ. After we were satisfied with the appearance of the clay organ model, we let it air-dry to harden.

After a few days, the clay models would completely dry, and we could paint them with acrylic paint. We mixed several colors to get the perfect shade and painted in several coats to make it fully opaque. When the paint dried, we covered the models with a thin coat of sealant to prevent peeling and fading.

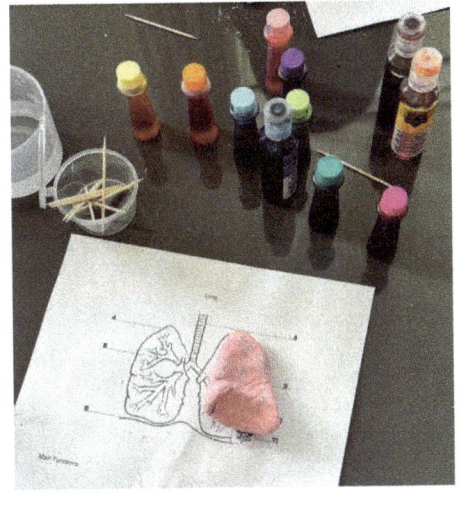

When the clay models were finished, we had to add them to the book. First, we would take pictures of the organ models and slightly edit them in Photoshop before inserting them into the book. Then we would research extensively about the organs' form and functions and write a description of them, along with an interesting fact. We also asked physicians if they had any helpful tips to protect each organ.

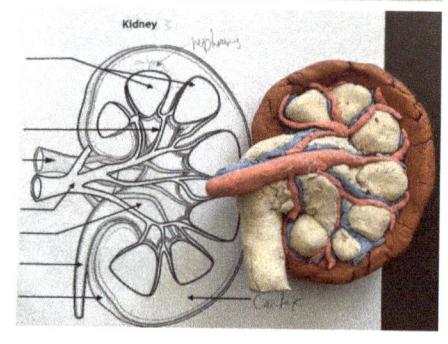

Finally, we would add personalized backgrounds and illustrations to each page. For example, we drew small figures to go with the fun facts and reminders to stay healthy. We also added the cartoon heart and brain characters throughout the book. Afterward, all that remained was a final review of the pages. We enjoyed every step of the process, and we are so happy to present you with the finished product.